This book belongs to

_____

Rylee's Rover Goes to Mars
All Rights Reserved
Text copyright © 2023 by JoAnn M. Dickinson
Illustrations copyright © 2023 by Daria Shamolina

www.JoAnnMDickinsonAuthor.com

ISBN: Hardcover 979-8-9855605-8-9
Paperback 979-8-9855605-9-6

Two Sweet Peas Publishing

Rylee's Rover

Goes to MARS

written by
JoAnn M. Dickinson

illustrated by
Daria Shamolina

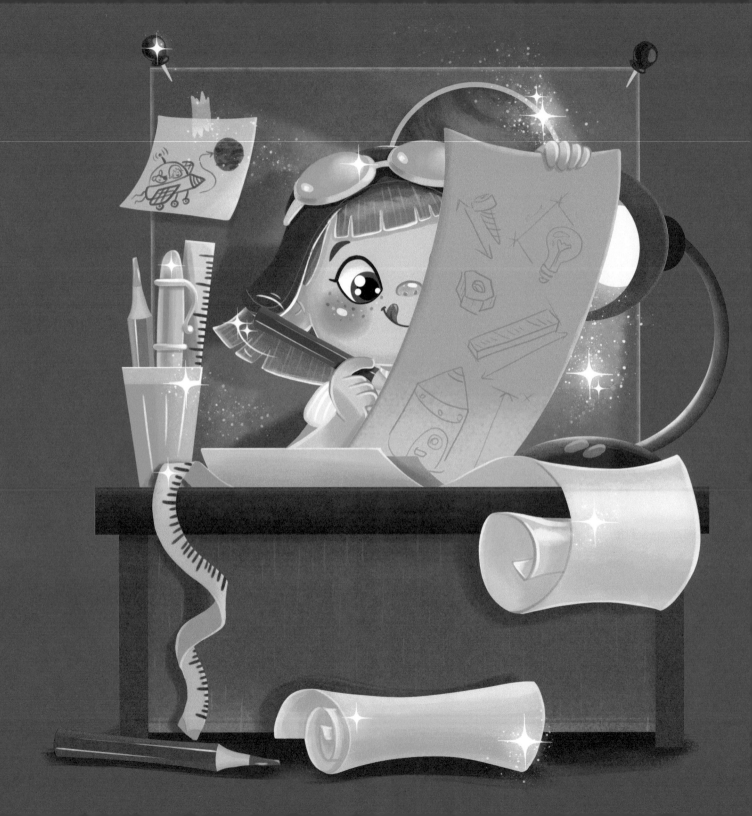

Young Rylee is reaching again
for the **stars!**
She has thought of a way
to explore **Planet Mars.**

She'll design a new **Rover**
with blueprints and schemes—
and then travel through **space**
as she follows her dreams.

She tells **Cosmo**,
"Let's travel to Mars very soon.
Help me build a new Rover
to fly past the **moon!**"

"We will **build** him so sturdy—
my Randal the Rover.
Recording with photos,
he'll **travel** all over."

"We'll go on a **journey!**
Let's 'give it a shot.'
We could have fun on Mars
while we're **learning** a lot!"

"Let's begin by **collecting** the most-needed parts, so we're building a Rover that **prints** perfect charts."

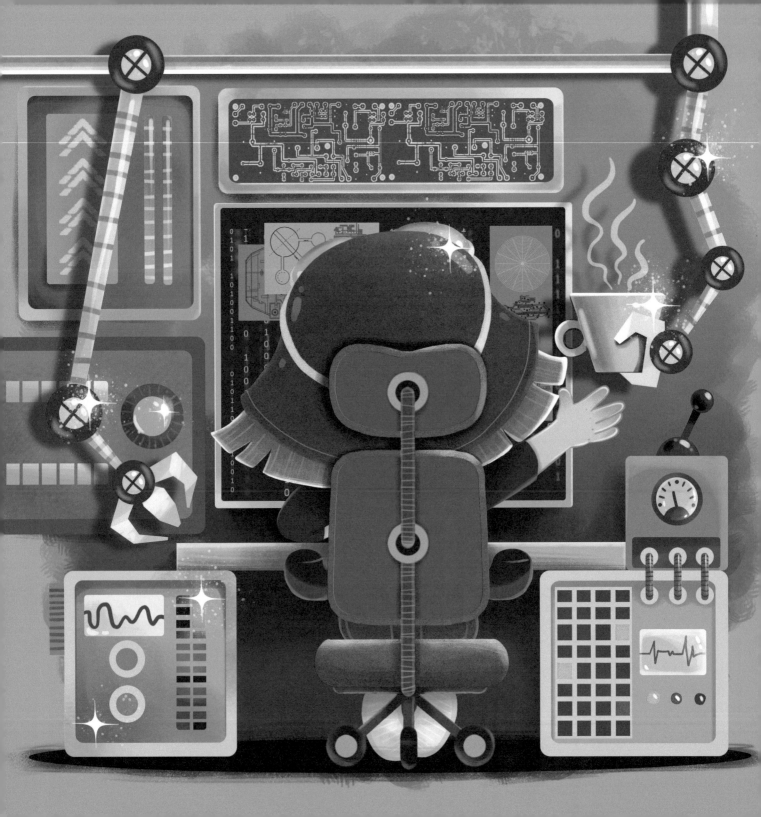

"As a skilled rocketeer
with the best of all rockets,
I'll gladly take care of
the engines and sprockets."

"Spectrometers made to
test **light**, we should take.
Those are brilliant **inventions**
we won't have to make!"

"Next, we'll need a few **covers**
to shield the computer—
Then, join complex parts
to resemble a **scooter**."

"We'll get **lights** and **wire**
for instrument panels
to carry fast **signals**
and switch between channels."

"Our Rover will need sturdy
wheels to go travel
on hills and deep sand—
and a lot of rough **gravel.**"

"Four cameras will **cover** both sides: Front and back. And the **mast** will be human-like—Painted jet-black."

"When we **start up** the rocket with Randal behind, we'll approach Planet Mars to see what we can **find**."

"We will learn by collecting
rare rocks and hard soil
in lots of containers
with **lids** made of foil."

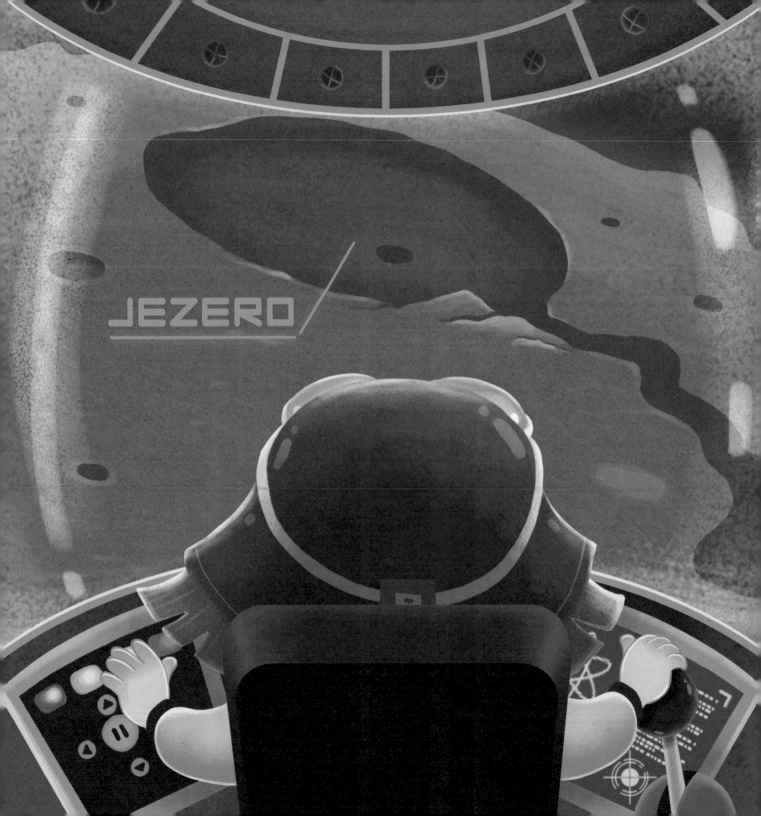

"Let's go near Jezero—
View craters and dunes . . .
and see Phobos and Deimos:
Irregular **moons.**"

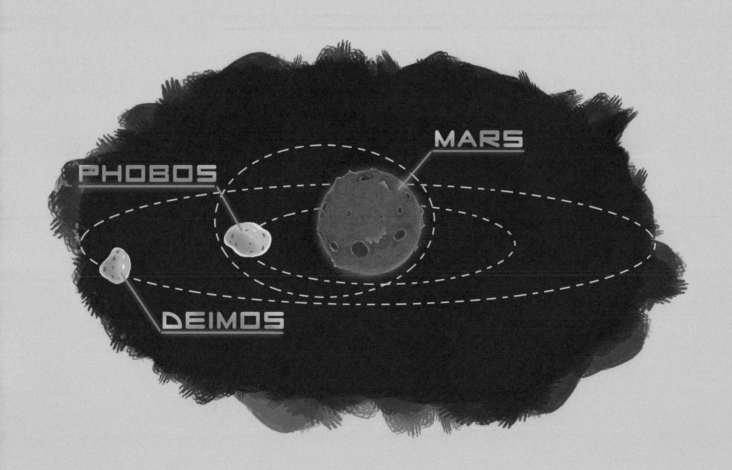

"We may see five-eyed monsters
with many an arm
building big **Martian** rock
farms with no fear of harm."

"Returning to Earth,
we'll enjoy the sky's glow...
and keep dreaming—while feeling
our planet's great flow."

"From the Earth, we'll see
Mars with our own naked eye,
as that planet shines bright in
the **dreamy**, big sky."

"There is much to discover.
Let's bring every friend.
And I hope that our fun
with **exploring** won't end!"

JoAnn M. Dickinson is a highly accomplished award-winning and best-selling author known for her captivating storytelling.

With eight self-published books already, she continues to push creative boundaries. JoAnn is expanding her popular series, including the *Lou's Zoo* Series, *Young Rylee* Series, and a new *National Park* Series. She invites readers on adventures of unforgettable, imaginary experiences.

Follow JoAnn's author journey at www.JoAnnMDickinsonAuthor.com

Other books available:

Daria Shamolina began her illustrating adventure at the age of **14** when she was hired for her first job at a newspaper created by teenagers. She later studied at her local University and began professionally publishing. She has illustrated multiple children's books and has a special talent for creating adorable, colorful and bright characters.

Would you like to schedule an author visit with

JoAnn M. Dickinson

Ask your teacher or librarian to email her:

JoAnn@JoAnnMDickinsonAuthor.com

Young Rylee Series

JoAnn M. Dickinson
Author.com

Made in United States
Troutdale, OR
12/11/2023

15714161R00026